The Origin and Decline of the

'Ice Ages'

The Real Cause of the 'Ice-Ages' A Mystery No Longer!

By Alan Dover

Also published by this author:

The Dynamics of Spiral Planetary Motion

www.onplanetarymotion.com

ISBN-10: 1543073239
ISBN-13: 978-1543073232

DEDICATION

To the current vital, heroic Syrian peoples struggle opposing the continuation of proxy wars and to maintain the Right of Nations to Self-determination, with freedom from foreign interference

The wonder of the Spiral
natures perfect mode of motion

Long obscured by men,
with self-interest their devotion.

They hold it concealed as long as they can,
with a medieval notion,
of a conjured up 'Wobbling' Earth,
a mere salve, a perpetual lotion.

Now, the spiral mode is bursting free
from their devious mathematical potion.

Table of Contents

ACKNOWLEDGMENTS

To my sons and daughter for their valuable assistance in proofing formatting and publication of this present work and also the general support of my whole family over many years.

To Stev Ominski for his enthusiasm and advice on front page editing of his wonderful artwork depicting ice age flooding along latitude 45° North.

The artist's description of "Ages' End":

From a suite of work about the Missoula Ice Age Floods created by painter Stev H. Ominski

"We are standing upon the site of the present day Vista House (Oregon), on a rocky promontory 733 feet high above the river, in The Columbia River Gorge. You, the viewer are looking east at the approaching waters and ice and debris that will soon back up behind the natural hydraulic constriction of this relatively narrow portion of the gorge, causing the eventual topping and inundation of our vantage point. Imagine the thundering sound, the rush of the change in air pressure, the wind, and the birds and animals fleeing past you in alarm." In this particular Ice Age, the event illustrated repeated itself nearly a hundred times.

"Ages' End" a 24"x 48" acrylic on panel was produced in 2006.

Preface

Spherical Motion

Spherical Motion as depicted here reflects planetary motion
in all its simplicity. It is governed by the law of gravity
which produces Spiral motion. This law applies to all
spheres in nature. Since all spheres in nature have two
forms of motion, spin and orbit, from the stars down to
atoms. They obey the same laws of motion a movement
about and towards a centre, the mode of development
throughout the solar system. As Dr Robert Hooke pointed
out 1674 in his work he proffered to Isaac Newton. 'An
Attempt to Prove the Motion of the Earth From
Observations:

> "....but it is a notion, which if fully prosecuted as it
> ought to be, will mightily assist the Astronomer to
> reduce all the Coelestial Motions to a certain rule, which
> doubt will never be done true without it. He that
> understands the nature of the Circular Pendulum and
> Circular Motion, will easily understand the whole ground
> of this Principle, and will know where to find direction
> in Nature for the true stating thereof. ....But this I
> durst promise the Undertaker, that he will find all the
> great Motions of the World to be influenced by this
> Principle, and that the true understanding thereof will be
> the true perfection of Astronomy."

Dr Robert Hooke   'Attempt to Prove the Motions of
the Earth from Observations'. 1674.

As we can see here displayed in the solar system, the contradiction between the two forms of motion of the spheres is a life-giving source. The Solar system, like all other things in nature comes into being and goes out of being. As we know, life is only temporary and relative. Fortunately we understand as a wise man once said, 'there is an absolute within the relative'. So relatively speaking, human life and society is an absolute within the relative, here today but gone tomorrow, a natural law.

Since all things in nature have their opposites, motion itself, elsewhere in the wider universe, there can be a movement about and away from a centre. The sphere observed in our universe is nature's simplest form of matter and therefore its motion is the simplest form of motion, the Spiral.

Alan Dover

Chapter 1: Introduction

Earth's Slowly Changing Inclination

The question has to be asked, why is the 'Ice Age' still a mystery? The answer is simple enough; it is still a 'mystery' because of man's *self interest*, which diverted him from the truth.

My first discovery in Astronomy has been the discovery of a great deception of man by man, corrupting the science of astronomy. Whereas, when many years ago I *independently*, looked nature in the eye, nature responded with the simple truth, the *spiral*. I had no reason to spurn it, like so many professionals have done. To me, it was a gift, a key to unlock the mystery of our solar system and more. It was an honor to accept that key, I was free to accept it because I had no invested interest in supporting a lie. Without realizing it, the spiral key was to eventually open before me the prize, the whole process of the ice ages. The cause of the 'ice ages' has long been a mystery, with it, are the mysteries of evidence of forests being revealed from under melting ice in near north polar latitudes. Also of snap frozen Mammoths being found from beneath receding ice fields and frozen Tundra and more subtle mysteries besides. All these are revealed when the real cause of the ice age is understood. . All the pieces come together like a jig -saw puzzle.

The reader will find that the whole process of the ice age is the product of Earth's very slow but continuous change in its inclination to the Sun. A full explanation of this change of obliquity is described in my earlier work on 'The Dynamics of Spiral Planetary Motion' that gives rise to the discovery of the real cause of the 'ice ages'. As a consequence this work is presented on the presumption that the reader is or will become, familiar with the recorded facts concerning this amazing phenomena, caused by Earth's continuously changing inclination. Known as the obliquity of the ecliptic. ε^{1}

The Spiral - the Cause of the Ice - Age

Spiral planetary motion has really been neglected by science, its true motion has not been revealed. It has not been recognised that a spiral orbit performs two functions with each complete orbit. One function of the spiral annually carries the Earth axis for example, a little bit closer to the Sun. as shown in **Fig. A**. a plan view of the pole, from where the **movement** toward a centre is recorded, annually **as** 0".47 annually.

---

[1] The astronomical symbol for the obliquity of the ecliptic = Ɛ

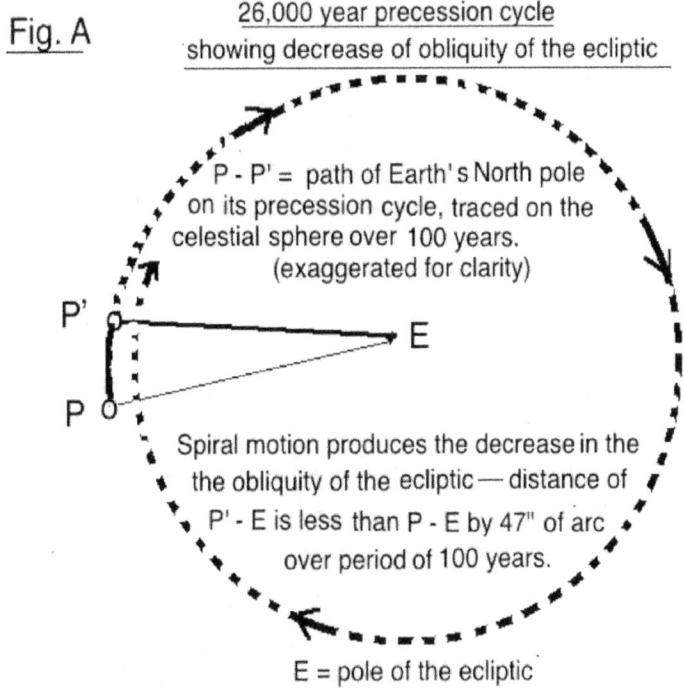

Fig. A

26,000 year precession cycle
showing decrease of obliquity of the ecliptic

P - P' = path of Earth's North pole
on its precession cycle, traced on the
celestial sphere over 100 years.
(exaggerated for clarity)

P'

E

P

Spiral motion produces the decrease in the
the obliquity of the ecliptic — distance of
P' - E is less than P - E by 47" of arc
over period of 100 years.

E = pole of the ecliptic

At the same time **Fig B** we can see that the spiral in one complete orbit also progresses along a path that carries the pole about the precession cycle at the rate of 50".2 annually. As a result of the combined function we perceive the movement of the axis towards the centre tells us that the equatorial plane with each orbit, simultaneously moves the inclination angle slightly towards alignment with the plane of the ecliptic. The spiral movement is not a simple single function it has a dual function.

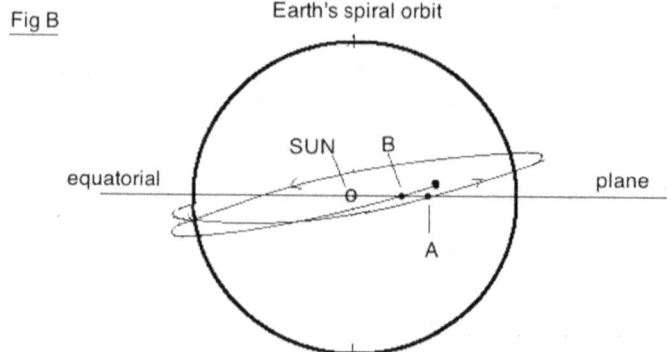Alan Dover

The precession of the equinoxes advancing along the equatorial plane

These are the qualities of the spiral that drive the Earth and Solar system, a continual progressive movement

It is known that the inner satellites of the planets are orbiting close to their planets equatorial plane. It is also recognised by many astronomers that a satellite, while orbiting, passing above and below the equatorial plane of its planet will have its tidal bulge moving continually up and then down. A stressful friction creating movement that progressively makes the satellite's orbit approach alignment with the equatorial plane of the planet. However, the actual dynamics of how, such an orbital movement accomplishes this has never been considered. But Isaac Newton's "bending moment" is a first step toward recognition of the spiral law of planetary-motion operating between a satellite and its primary.

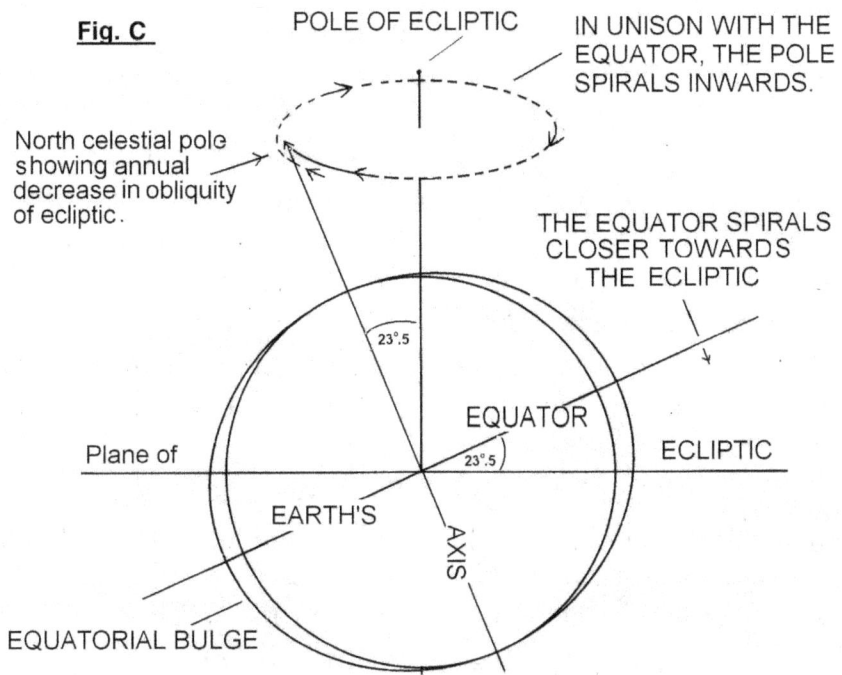

Fig. C

POLE OF ECLIPTIC

IN UNISON WITH THE EQUATOR, THE POLE SPIRALS INWARDS.

North celestial pole showing annual decrease in obliquity of ecliptic.

THE EQUATOR SPIRALS CLOSER TOWARDS THE ECLIPTIC

23°.5

EQUATOR

23°.5

ECLIPTIC

Plane of

EARTH'S

AXIS

EQUATORIAL BULGE

The Bending–Moment as Determined by Isaac Newton
Updated to show the Spiral Form of the Rotation

The popular traditional description of the precession motion is depicted in **Fig.C** with Earth's response, as a satellite, to the gravitational force of the Sun and Moon. "A bending moment", acting on the equatorial bulge of the Earth; which tends to rotate the equatorial plane into that of the ecliptic just as Newton indicated. It has never been investigated as to how the pole of diurnal rotation can really accomplish any change in angular distance from the pole of the Ecliptic. Although the annual decrease is officially recognised, the *spiral nature* of the movement of the axis decrease is not. The precession cycle is always depicted as a circle, a mechanical description.

In contrast we recognise that the decrease in the

obliquity is and always has been, as recorded, a continuous movement towards the ecliptic as Newton suggested. It is only through a spiral rotational motion that such a change of an axis can be accomplished. It follows that the poles will describe as they do, spiral paths on the celestial sphere.

Such a fundamental view of planetary motion opens the way for a better understanding of the obliquity of the ecliptic and of the climatic changes on Earth, in the past and the changes to come. In accord with the spiral law of planetary motion we are able to proceed to investigate Earth's travels in the past and into the future.

The bounty that presented itself was a reward indeed – the cause of a mighty ice age. The whole question of the ice age origin and demise rests on the law of spiral motion inherent in all planets. Understanding this spiral movement is the key to understanding the cause of *the ice age*.

To help understand the phenomena of the so called 'Ice Ages' I have set out a short series of images to simplify our understanding of what is a natural process in the changing balance between water and ice on our planet. This balance is not fixed in perpetuity and it like all things in nature, has a beginning and end. The dynamics that drive this repeated transfer from water to ice and back again is dependent on two factors Earth's orbit and its spin. The first gives us our daily tempo which is a varying balance between the length of night and day. The second is Earth's orbit about the Sun that produces continuous brightly lit hemisphere and a continuously shaded or dark hemisphere rotating around Earth annually. The axial inclination to the sun gives us our annual cycle of seasons, spring summer autumn and winter.

There is a conflict between our daily rotation and our annual orbit, which is driven by the slow continuous[2] change in Earth's inclination to the Sun. This will be apparent when viewing the images presented here. The conflict is between the balance of night and day as Earth's inclination shifts, it will be seen that in some instances, hemispherical areas become continuously shaded (night) through much of the year. And alternatively other hemispherical areas become continuously sunlit through much of the year. Though this will seem very crazy and confusing, its simplicity will become apparent when viewing the cycle demonstrated by the images.

During this conflict the *annual* balance in the transition of water to ice and ice back to water, is a process that makes only imperceptible changes to sea level annually. Whereas, the factor that is responsible for the greater *long-term* changes in sea level governing the amount of growth and then elimination of ice on our planet, is the *continuous* reduction in Earth's inclination to the Sun. More ice, lower sea level – less ice, higher sea level, essentially a process governed by Earth's changing inclination to the Sun. This changing angle to the Sun is driven by spiral planetary motion and is fundamental to our existence, since it governs our slowly changing climate.

The images here portray the incredible small annual advance of Earth's axial inclination toward equating with the pole of the ecliptic. To understand the whole sequence

---

[2] See the publication 'The Dynamics of Spiral Planetary Motion' concerning the continuous decrease in the obliquity of the ecliptic. Available from Amazon books.

I present the process in a quadrant of the 360° compass, with Earth's passage from 90° obliquity through to 0° obliquity (from B to E as depicted in **Fig.B-8 chapter 4)**.

The hemisphere wide freezing and the violent, global and hemispherical massive 'ice age' freezing and flooding, is displayed herein. Subsequently affecting Earth's topography first globally and then hemi spherically in those latitudes 40° to 50° North and above. When compared with the geographical evidence being continuously and extensively gathered today concerning the retreat of the Cordilleran ice sheet from latitudes 40° to 50° North; as in the Northern states of the USA by such as the Ice Age Floods Institute

The evidence corresponds precisely with the second phase of the decline of the ice age. A protracted process of the reduction of ice to water resulting in higher sea levels. This evidence stretches around the globe on these latitudes and of course this process is mirrored in the southern hemisphere, although there is less land mass in the southern hemisphere to provide the strong evidence that exists as in the North above 45°latitude.

Of course the evidence of global icing and flooding is all around us, all the canyons around the world stand as monuments to its power and ferocity. It will be seen that as the Earth has proceeded earlier through quadrant 3, the difficulties of giving rise to and sustaining life in what would have been chaotic conditions. It must have been virtually impossible until such time that Earth reached the obliquity 45°, where equal day and night was becoming in part, more stabilized.

It will be realized that this interpretation of our planets motion is simply based on the 2000 year recorded annual 0".47 decrease in the obliquity of the ecliptic. This being a continual movement toward equating with the pole of the ecliptic and thus, moving toward equating with the Sun. The current and historical interpretation held by astronomy for perhaps 300 years, denies that that this movement of Earth's axial inclination is a continuous movement but is a mere wobble of Earth and that Earth's axis will not move closer to equate with the Sun as Sir Isaac Newton once inferred. The situation has, for so many years left the science of astronomy in the embarrassing situation that it is in, by promoting an interpretation of perpetual motion that is contrary to the principles of both the sciences of Astronomy and Physics. In short they have spurned nature's spiral.

Alan Dover

Chapter 2: The Ice Age Explained

The First Phase of the Decline of the Ice – Age

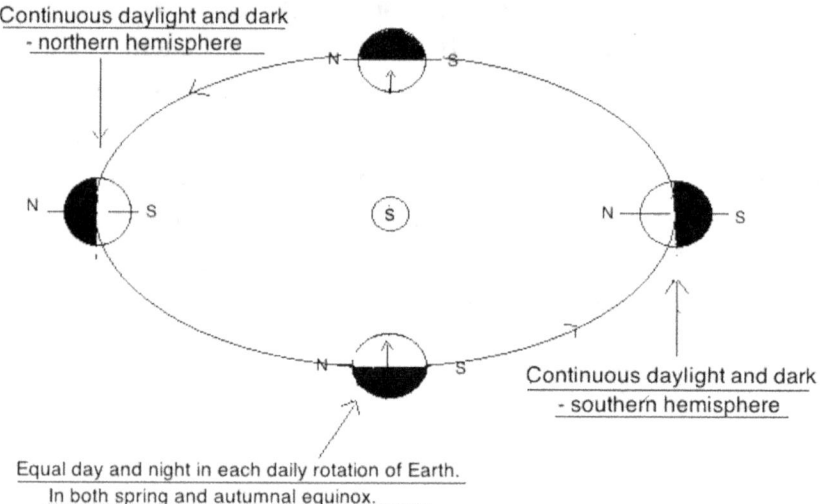

Fig. A-1 Daylight and darkness on Earth on each annual cycle about the Sun at 90° inclination

Continuous daylight and dark - northern hemisphere

Continuous daylight and dark - southern hemisphere

Equal day and night in each daily rotation of Earth. In both spring and autumnal equinox.

It is very important that the reader carefully examines and understands this following drawing and description of the basic, origin preceding decline of the ice age, commencing at 90° inclination of the Earth's axis. The ice ages are all about the changing angle of inclination of Earth's axis to the Sun.

For those who do not familiar with how the Earth changes its inclination, and then the reasons for the Earth's angle to change can refer to 'The Dynamics of Spiral Planetary Motion'. Where the facts are fully explained.

As an example, Fig A1 gives the extreme situation of inclination with the Earth's axis laying parallel with the ecliptic, 90° inclination to the pole of the ecliptic. This creates a bizarre climatic situation as can be seen above with Earth's rotation from East to West at the Winter and Summer solstices, it can be seen that half of the planet is receiving continuous sunshine and the other continuous darkness.

But of course, it will be noted that just 3 months later in its annual orbit the Earth arrives at the spring equinox where it will be experiencing equal day and equal nights. A seriously dramatic rapid change in climate to say the least. This will of course be repeated at each solstice annually.

This 90° angle of inclination is the apex of the ice age, from which the ice age begins its decline then we will examine the decline through to the present day 23° and beyond.

The Earth in its orbit always has a shaded hemisphere and a sunlit hemisphere that travels about Earth in sync with its orbit we take for granted. This is a stable basic condition that gives us night and day but we need to look closer to appreciate it better. As illustrated, the passage of Earth through each of the four phases takes 90 days. But this is all dependent on Earth's inclination to the Sun that has historically been steadily changing, thus transforming Earth's climate, as we shall discover.

It will be noted in the above Figure **at 90°ε** for example,

15

at the winter solstice the shadow covers a whole hemisphere and when it 'orbits' around anti – clockwise by 90° into the spring equinox the shadow still covers a hemisphere of 180°. We just need to be aware that, at the solstices one hemisphere is completely dark and the other completely sunlit for 90 days whilst as the shadow passes through the equinoxes it is *always* equal day and night at as displayed. This is due to the fact that, as illustrated the direction of the daily rotation of the Earth becomes the dominant factor that makes this possible.

Of course the leading and trailing edges or terminal points of the shadow one end creating and the other end releasing a frozen area from under the shadow. This means that any given point on Earth travels through each of the four 90° quadrants, a cycle of prolonged 90 days freezing and then 90 prolonged equal day and night defrosting. A cycle which has for 90 days been frozen and now exposed will rapidly defrost in 90 days. Twice a year, a mighty freezing and melting engine

At present the Earth's spin creates an axis that has an inclination to the Sun that gives us our seasons. Our investigation now is focused on the progressively changing inclination of the planet Earth to the Sun. The current scientific establishment does not favour the progressively annual changing aspect of Earth's inclination to the Sun that nature exhibits and so has chosen to simply ignore it in favour of a more acceptable 'invariable' explanation.

Be that as it may, we however heed nature's law of change and pursue the possible changes in Earth's inclination and to do this we must first acknowledge nature's spiral law of planetary motion. So here, our first step is to recognize this spiral motion of our planet. This will then take us back on an amazing journey through millions of years of Earth's development to a time when no life could exist, as we know it.

The simplest way I have found to explain the progressive and continuous change in Earth's inclination to the Sun, is as follows.

We commence by projecting ourselves back through millions of years of possible inclination of Earth to the pole of the ecliptic as depicted here in Fig.A-2. commencing at A with Earth's axis lying parallel to the ecliptic at 90° to the obliquity of the ecliptic (ε). Then later with its axis reduced to 45° at B then 23° at C.

The indications over two thousand years are that the Earth is annually closing the obliquity of the ecliptic. So it is obvious that the Earth has a past history that is directly related to a continuously changing inclination. In due course you will see that this not an unlikely situation when you see the dynamics of the ice age emerging through earlier inclinations - 90°, 45° and 23°, with the phenomenal transition of ice to water and back again to ice; twice a year! A fantastic display of the power of water to ice and back again due of course to sunlight, darkness and the spiral orbit.

Fig.A- 2. is a guide to the sequence of the diminution, or *decline* of the ice age. We point out here something of importance, since our focus is on the decline of the ice age taking place down from 90°. We note that the *rise* of the ice age preceded its *demise* and reached its peak at 90°, being a reverse of the decline. It historically, precedes the sequence that is depicted in Fig.A-2, the starting point for our investigation the demise. From 90° to 23° and beyond.

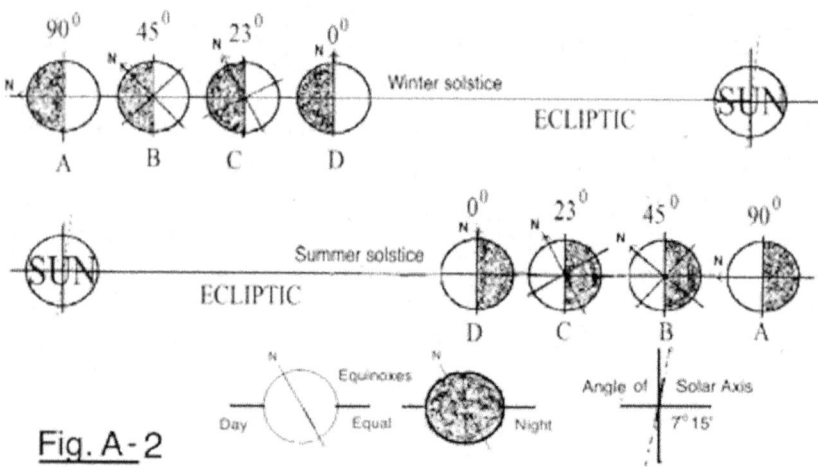

Fig. A-2

All the following images demonstrate the two conflicting motions of our planet, the annual motion and the daily motion, (orbit and spin) through several different inclinations, from Fig.B-1 to Fig B-8

Starting with the inclination at 90° (refer to Fig.A2 at 'A') we view in some detail, in Fig.B-1, the two motions of our planet's spin and orbit and their effect on each other. The process displayed in **B.1** reveals the dependence of Earth's climate on the angle of the obliquity at any given time. We will then examine the continuation of the process as Earth advances to decrease the obliquity of the ecliptic (ε) at several other inclinations in its progress.

Two Inter-related Features of the Ice Age

The two inter-related features of the ice age are the rapid icing and the rapid melting taking place, each over two periods of 90 days; that being the rate of the orbit and thus the rate of the shadow passing around the Earth. There are the two stages   in the orbital cycle, one based on the Spring equinox and the other on the autumnal equinox. This completes a Bi-annual cycle of freezing and melting, one in the Northern hemisphere and one in the Southern hemisphere.

The leading edge of the shadow in the 90-degree inclination for example, is of course responsible for the rapid freezing action and in this inclination freezing has its greatest coverage declining in its intensity as the inclination changes.

The motion of the Earth's shadow never changes on its annual path about the planet so has no direct influence on the changes in Earth's climate. However, Earth's ever decreasing inclination to the Sun is wholly responsible for the constant change in climate, over millions of years.

The freezing cycle.

The trailing edge of the shadow is releasing the frozen area to the rays of the Sun hence the flooding action. This too has its greatest intensity at the earlier stages declining as the inclination changes.

It will be seen that at the peak of the ice age, both the Northern and Southern *hemispheres* were *annually* completely frozen over in turn, for ninety days each. This has left evidence of ice covering and flooding of the whole *sphere*, which gives rise to the notion that the whole sphere was for a long period was once covered in ice. For example the 'Snowball Earth' theory; it is true that collectively the two constituted global ice coverage in the first stage, each for 90 days But annually the cycle was punctuated twice by the two equinoctials of equal day and night, also amounting to 180 days. So it can be seen that it was never a question of a completely frozen globe.

Arising from this amazing *annual* phenomena of countless cycles that we have called the 'ice ages' and because of the evidence of numerous cycles, it has been thought that there have been numerous ice ages. This is not the case; there is only evidence of one protracted mighty ice age.

We come to the other feature of the ice age; the melting or more correctly the flooding cycle. Just as the ice was forming hemi spherically in the terminal leading edge of Earths shadow in its terminal wake, as it moved on into the equinox, any point on the frozen surface was next progressively exposed to full rays of the Sun for 90 days. This would have consisted of a wave of mainly fresh water washing over each hemisphere North and South in turn, each year. This must have been the cause of the enormous flooding accounting for the creation of the gigantic canyons over all continents of the world, especially at that peak period of 90°, over many thousands of years.

It will be perceived that the angle of canyon walls is

governed by the steady rate of decline of the quantity of floodwaters from broad water flows at the start, to become progressively narrower as the volume reduces.

Following the peak period of the ice age 90° ε the following figures demonstrate the slow decline of ice from planet Earth.

Fig B-1 90° inclination shows shadow at maximum of ice creation and lowest sea level with inclination at 90°. It will be noted that any point on the northern half of the planet is in darkness for 90 days continuous and any point on the southern half of the planet is in daylight for 90 days continuous. This is of course the extreme effect of climate change.

During this period of 90° inclination (possibly many millions of years) a freezing temperature producing a vast area, a whole hemisphere of ice. But this enormous amount of ice would be spread relatively thinly because it was so rapidly formed and accumulated in a mere 90 days, it could equally be melted in 90 days. A delicate balance subject to continuous change nevertheless.

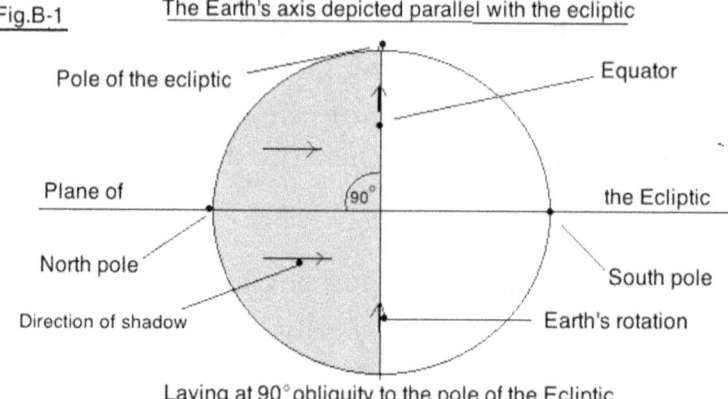

Fig.B-1   The Earth's axis depicted parallel with the ecliptic

Laying at 90° obliquity to the pole of the Ecliptic

Next comes the most dramatic rapid change of events as the shadow enters on its annual passage through the equinoctial.

Earth's *daily* rotation now comes into effect. The shadow's freezing effect being rapidly eliminated through its subjection to 90 days of exposure to equal days and nights when it proceeds into and through the next quadrant, the spring equinox. At the same time at this juncture the shadows trailing edge will be uncovering the frozen surface exposing it to the Sun, thus melting, as 12 hours of Sunshine during the equinox daily causing the advancing flooding where we see the equinoctial hemisphere becomes completely ice free, as in Fig.B-2

This passage of the shadow through the spring equinox completely halts all ice formation at its leading edge, whilst rapid this *global flooding* takes place on its trailing edge to advance across a hemispherical wide front, from the North pole to the equator. This process is a bi-annual event, the first stage of the *annual* cycle, the second stage will be similarly repeated at the autumnal equinox, giving two massive *global* floods per year. The whole globe experiences flooding in the course of a year. Hence the evidence of canyons distributed globally gives the false impression of an ice bound globe at some time in the past. Another more obvious point of interest is that the area of melt is widely distributed along the melt front. As it advances it will cut channels in depressions which in the course of many years become the natural water courses along which the flow follows, becoming the canyons and of course this channeling of the widely distributed melt water will cause deeper and faster flows of the water along these courses than would the original widely distributed flows This would account for some estimates of water flows through some canyons being at very high speeds. It is notable that the most impressive canyons are found on the basalt plateaus, being of volcanic origin and of a softer sandstone texture than the harder rocky strata beneath. Like the Grand Canyons in the USA in a desert region they are particularly beautiful since they are mostly devoid of vegetation unlike many of the numerous canyons around the world.

Still looking at a decrease of ε of 90° inclination, during the shadows passage through the spring equinox, it is limited to nighttime 12 hours only. Displaying a hemisphere with equal day and night, the shadow will move on to enter the next quadrant where, free from equal day and night exposure, it will enter into the second freezing stage of the cycle during its passage, as at 'A' through the summer *solstice* in Fig. A.2.

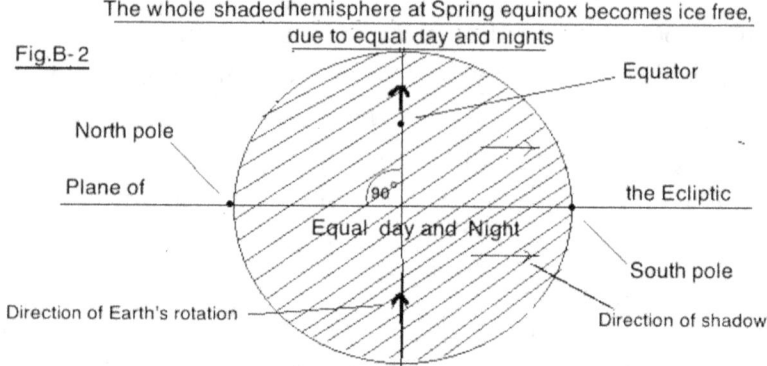

Fig.B-2

The whole shaded hemisphere at Spring equinox becomes ice free, due to equal day and nights

Equator

North pole

Plane of

90°

the Ecliptic

Equal day and Night

South pole

Direction of Earth's rotation

Direction of shadow

The Earth's shadow is about to pass into the next Quadrant

All Ice has melted as the shadow moves on to the next quadrant. That concludes the examination the first stage of the 90° cycle, the second stage is a repetition of the first, the shadow passing on to freeze and then through to melt in the *autumnal equinox*; completing the cycle. That ends the 90°cycle.

The two interacting motions, Spin and orbit can be viewed as follows – shadow advances around the Earth along the ecliptic at annual rate of orbit = 1° per day. Daily rotation, turning at 90° to the plane of the ecliptic produces equal day and night only at the equinoctials.

10° change in inclination

Moving on to a new situation, though the orbital cycle of the shadow is unchanged the Earth's inclination has moved as in **Fig. B.3**, this shows the effect of the reduction of Earth's inclination to 80° angle. It begins to get more interesting when we examine this first step in the reduction of ice and the increase in sea level. Though many thousands of years are involved in such a movement of 10° of Earth's inclination.

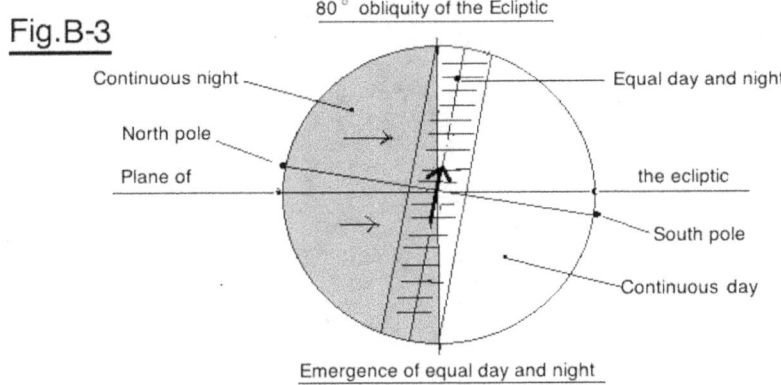

Fig.B-3

80° obliquity of the Ecliptic

Continuous night

Equal day and night

North pole

Plane of

the ecliptic

South pole

Continuous day

Emergence of equal day and night

We can see the broad front of the ice entering the spring equinoctial quadrant has already been reduced slightly and therefore the quantity of ice is already slightly reduced, due to the narrow strip of equal day and night appearing along the equator. There will be an accompanying slight increase of sea level. At the shadows trailing edge where the melting is taking place there is still clearly a global flooding front, though slightly reduced in its length.

The shadow will fully enter the next quadrant, the spring equinox and be subject to a halt in freezing with the accompanying rapid melting of ice; due to the 12-hour daylight limiting the shadow from freezing the area for 90 days. Then, will continuing on the cycle as before into the second stage of freezing and melting. That concludes examination of the 80° example.

We now select Fig.B-4. At 66°ε, another example of the effect of the decreasing angle of obliquity on the diminishing quantity of ice to be melted.

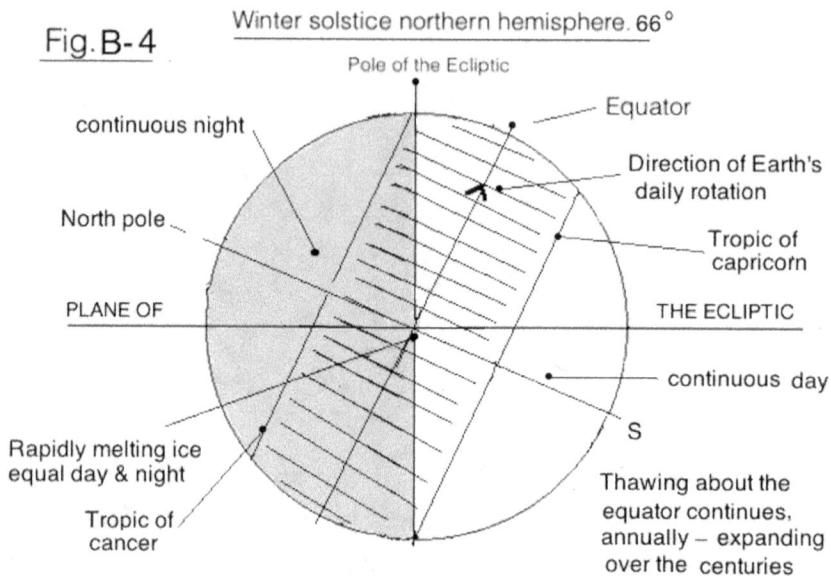

Fig. B-4

Winter solstice northern hemisphere. 66°

Pole of the Ecliptic

continuous night

Equator

Direction of Earth's daily rotation

North pole

Tropic of capricorn

PLANE OF

THE ECLIPTIC

continuous day

S

Rapidly melting ice equal day & night

Thawing about the equator continues, annually – expanding over the centuries

Tropic of cancer

Earth's pole inclined to pole of the ecliptic 66°

Here in **Fig.B-4** we see a significant change. While the flooding front has now been even more reduced and we notice the tendency for the ice forming front of the shadow to be retreating toward the North polar region in Earth's winter solstice. The trailing flooding, now is somewhat less than a global front. Likewise, as before, the shadow moves into the equinoctial quadrant to totally halt it's freezing capability and all ice to be totally melted. Then again, proceeding to the next quadrant where large sections of the polar latitudes of the planet are in either continuous night or dark quantities.

In the remarkable next decrease in inclination, 45° we

can consider that there is more likely to be suitable conditions for supporting forms of life appearing in those equatorial latitudes, with a consistent range of the length of equal day and night, though it is only equal on a narrow band about the equator. There are wider temperate zones covering areas in latitudes moving away from the equator. Thus we enter the distinct second phase of the ice age.

CHAPTER 3: The Second Phase

The Tipping Point - 45°

Here, in Fig.B-5 we can see that the movement from 90° to 45° inclination constitutes the end of the first phase of the diminution of the ice age and ice age flooding. It is in fact the *tipping point* where the change from global flooding to hemispherical flooding occurs.

From this inclination onward, we see a marked change in Earth's climate with the *creation of two ice caps* and a large region between latitudes 45° North and 45° South of having suitable conditions for some development of life forms, especially around the equatorial regions.

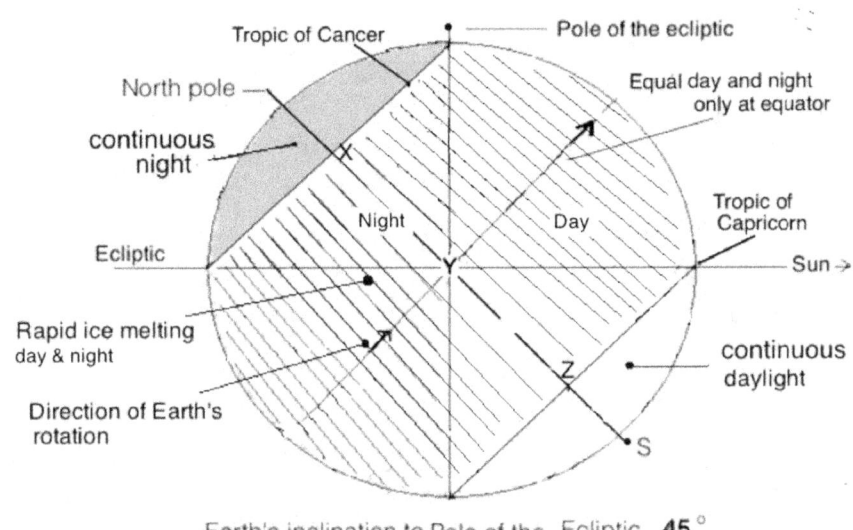

Fig.B 5 At Northern winter solstice - hemisphere showing variation from periods of continuous night & day to equal day and night

Tropic of Cancer
Pole of the ecliptic
North pole
Equal day and night only at equator
continuous night
X
Night
Day
Tropic of Capricorn
Ecliptic
Y
Sun →
Rapid ice melting day & night
Z
continuous daylight
Direction of Earth's rotation
S

Earth's inclination to Pole of the Ecliptic **45**°

While at the regions above latitudes 45° North and 45° South, the shadows effect would still subject them to the bi-annual freeze and melt with the accompanying now *hemispherical flooding.*

This bi - annual freeze and flood would as can be seen, sweep across the whole Northern hemisphere wide as the then tropic of cancer. We can now see an immense area between latitudes 45° North and 45° South as the leading edge of the shadow passes into the equinoctial transforming its effect into 12 hours of night and 12 hours of day. It seems fairly certain that at this time life forms are becoming established but still subjected to flooding at the higher latitudes.

The Emergence of Two Ice Caps

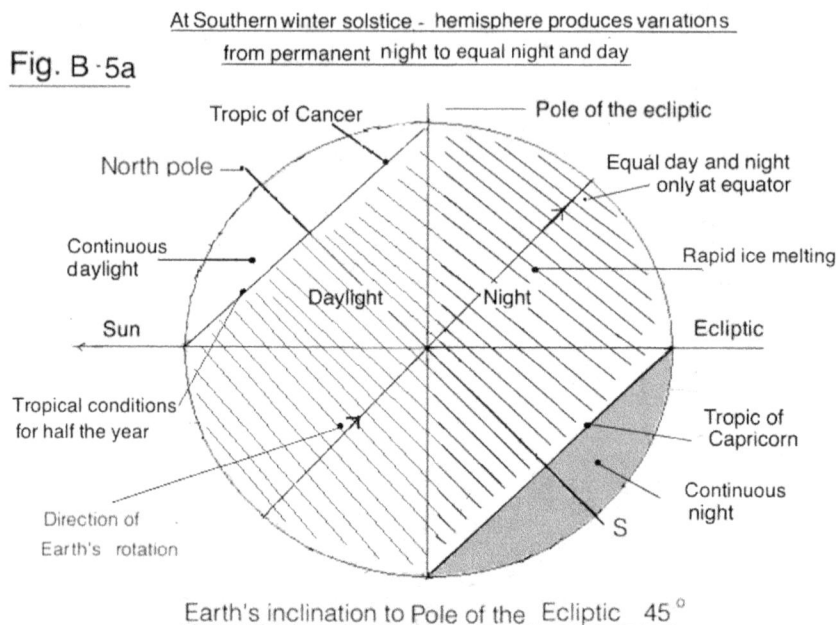

Fig. B·5a

At Southern winter solstice - hemisphere produces variations from permanent night to equal night and day

Earth's inclination to Pole of the Ecliptic 45°

We are aware of the mirror image that takes place continuously in each stage of the cycle where annually there will be a similar ice cap appearing at its Southern pole, as shown here in **Fig. B-5a** at 45° South.

Now in **Fig.B-6** we can really begin to see the facts of the matter, especially as we continue on through the second phase to our present day inclination 23.5° with our moderate seasonal changes each year with our docile ice caps with no extensive freezing or flooding. The ice is diminishing and the sea levels are rising. A terrific amount of ice has been reduced to water. But today only a relatively small amount of ice to water and back again still takes place twice a year. Once at each pole, we are in the passive era of the ice age.

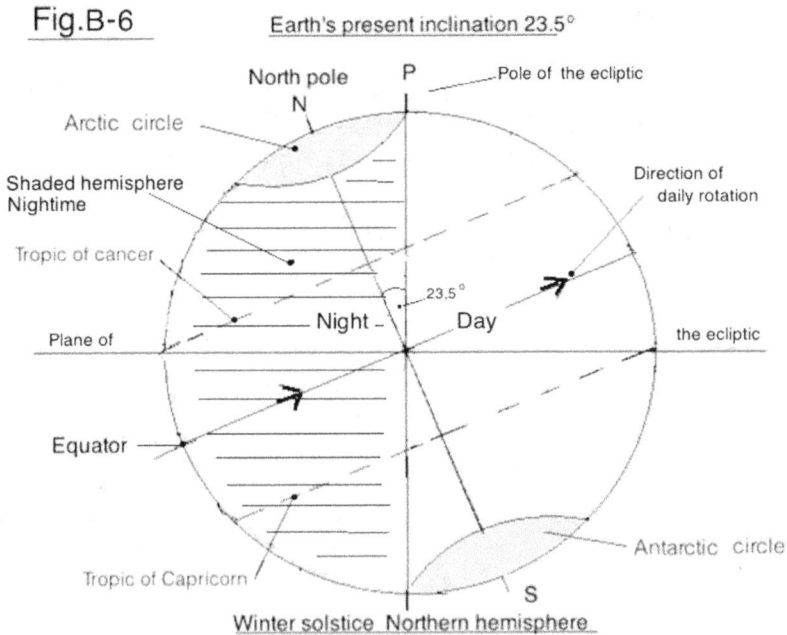

Fig.B-6   Earth's present inclination 23.5°

Moving on now to the future in Fig.B-7 we can view the likely situation with regard to Earth's climate and consider what a fantastic natural wonder has preceded mankind's existence here on Earth.

There is one more thing to consider though and that is As Earth's inclination closes with the pole of the ecliptic, we are aware that meanwhile, the path of the Earth around the ecliptic has itself, due to its spiral motion, been very slowly moving closer to alignment with the solar equatorial plane. Therefore, by the time the Earth's axis eventually aligns with the pole of the ecliptic, then the   pole of the ecliptic would be very close if not already aligned, with the Solar axis bringing all three axes to alignment, as depicted in **Fig B-7.**

At zero degrees inclination there are no longer any seasons and ice will have been all but eliminated except for higher altitudes of mountainous regions. The sea level would naturally be at its maximum and the equatorial region would be a very hot and dry arid zone.

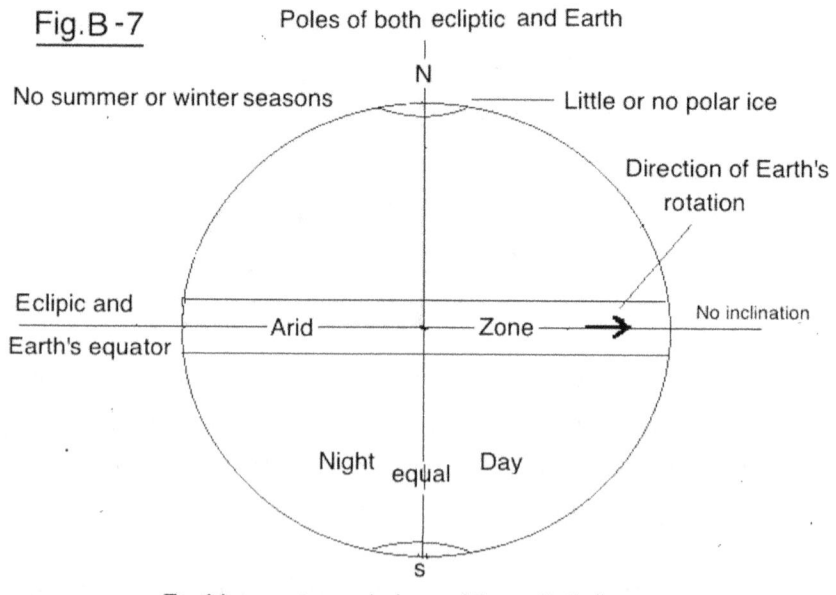

Fig.B-7   Poles of both ecliptic and Earth

Earth's equator and plane of the ecliptic in sync

Comparing Precession Cycles

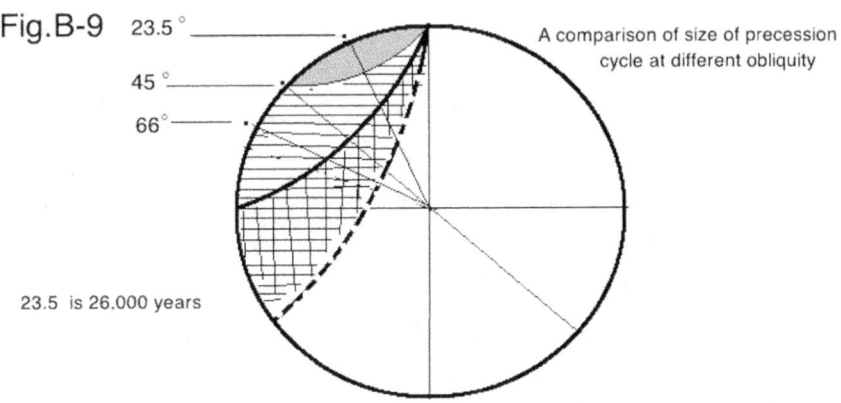

Fig.B-9   23.5°
           45°
           66°

A comparison of size of precession
cycle at different obliquity

23.5 is 26,000 years

Fig.B-9 shows an interesting point about the Precession cycle traced on the celestial sphere. It will be noticed that the precession cycle has been reducing in size where today it is estimated to be a cycle that takes almost 26,000 years and as the obliquity continues to decrease (due to an increase in the rate of decrease)[3] it will eventually cease altogether at 0°. We will wonder how much longer in years, the precession cycle would have been taking at the earlier inclinations, with the cycle's greater circumferences?

---

[3] Lieut. Col Drayson, R.A. F.R.A.S.

CHAPTER 4: A Greater Ice – Age

The Origin and Growth

The process of the *decrease* in the amount of ice on planet Earth should not be seen in isolation. This study seems to indicate that the decrease in ice that we are currently experiencing is only half of the phenomena. It may be better understood when looking at the spiral progression of Earth's orbit about the Sun as is displayed in **Fig. B-8**.

We can see that Earth is descending through quadrant 4 from 'B' through 'D' to 'E' and as our Figures reveal that the process is one of the dissolution of ice from its maximum quantity at 'B' to a much lesser quantity today. Undoubtedly a process of diminution driven by changing *inclination* and daily increasing the exposure of Earth's entire surface to Sun.

The process of the decline of Ice on Earth's poles, causes us to consider that if the Earth had indeed earlier passed through quadrant 3, from 'A', rising to the apex of ice over water at 'B', then it is logical to assume that quadrant 3 is responsible for the *growth* of the ice over water ratio on Earth's passage to the apex. These are the facts indicated primarily by the recorded continuous natural annual decrease in the obliquity of the ecliptic, and also the measured and recorded movements of our *plane of the ecliptic* that though minute, is not to be overlooked.

Fig. B-8 Arrows on orbital perimeter indicate direction of orbits progression, bringing Earth's orbit into alignment with equatorial plane of the Sun at 'F'

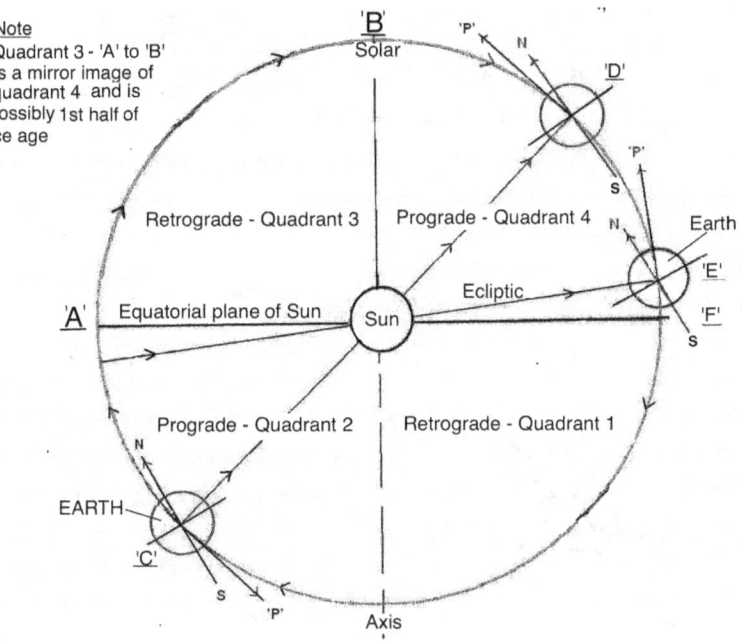

The most important of the ice age evidence that surrounds us, is regarded as a mystery, such as the cause of ice age flooding. Of which, the effect on and around Latitude 45° North is presently receiving a new found interest by geologists around the world. There are, as we know, many greater and lesser massive canyons on all continents and latitudes of the world. All are known to have been sculptured by water, all evidence of global flooding. Therefore geologists are currently seeking an explanation for hemispherical flooding in and around 45°latitude North. While there is still a need to explain the global flooding which is more ancient. But all are seriously stymied by the misinterpretations of planetary motion by astronomy.

The recognition of the spiral as the dynamism and mode of planetary motion is fundamental to understanding not only the mysteries of our planetary system but opens the way for a deeper understanding of nature throughout all our sciences. After all, the spiral is the natural product of gravity itself.

James Croll – Geological Evidence

The renowned geologist James Croll, summarized the situation in the **'Transactions of The Geological Society of Glasgow'**. Session 1866 – 67. Some excerpts from the evidence he collated in this work are as follows, titled:

'On the Change in the Obliquity of the Ecliptic; its Influence on the Climate of the Polar Regions, and the Level of the Sea.' (read April 18th 1867)

Quotations:

"….in the arctic regions, encrinites, corals and Mollusca belonging to the Silurian period, have been found in abundance proving that a warm sea must have prevailed in those regions, during at least a part of that age"

"It is well known that all over the arctic regions extending to the most northern limits that has yet been attained, coal and carboniferous limestone have been found in abundance."

"We have evidence of warm conditions of climate in North Greenland during Oolitic period. For example in Prince Patrick's Island at Wikie Point in Latitude 76° 20' N., and longitude 117° 20' W., Oolitic rocks containing an ammonite....and other shells of the Oolitic species were found by Captain McClintock."

" Sir E.Belcher found in Exmouth Island Lat, 77°16' North and long. 96 W.,

At an elevation of 570 feet above the level of the Sea, bones that were examined by professor Owen and were pronounced to be those of the ichthyosaurus. Mr Salter remarks that at the time that these fossils were deposited "a condition of climate something like that of our own shores was prevailing in latitudes not far short of 80° North " We know," say's Sir Charles Lyell "that Greenland was not always covered in ice for when we examine the tertiary strata of Disco island.

The list goes on and on, Atanekerdluk, Bell Sound, in Spitsbergen on 78° N. we find the Beech, plane Hazel and some species identical with those from Greenland.

"Again during the Pliocene period, we have well known glacial epoch with the Northern hemisphere to considerable low latitudes enveloped in one general capping of ice ....at a period somewhat later huge forests flourished in North Greenland and the regions about Melville Island, where at present not a shrub can grow."

More examples are given - of forestation in Wellington Sound lat. 75° 32' North.

Of the remains of an ancient forest in Banks Land Lat. 74°48' North'.

End of quotes.

The above evidence is impressive and interesting for it is clearly revealing a changing inclination in the North polar region whereby, a temperate climate existed *prior* to the climate changes to inundate the region with ice. This seems to indicate a period when the *ice ratio was rising*, sea levels were decreasing and the ice caps were advancing to cover land previously afforested exposed by earlier low sea level, in this first half of the ice age. This where the ice is advancing upward toward its apex at 90° where our decline was destined to begin.

Deprived of the truth concerning the continuous decrease in the Obliquity, which belies the annual change in inclination. James Croll, like many others, who followed; were frustrated in their quest by the misconceptions of astronomy concerning the annual decrease in the obliquity. They turned to the only inclination changes allowed by astronomy and eventually accepted the pathetic oscillating or wobbling Earth theory.

Ice Age Freezing and Flooding

The long list of abundant evidence provided by Mr. Croll points to the Northern hemisphere growth of the ice caps. As opposed to current retreat of the ice caps in the long term, the *whole process* is one of a cycle that converts Earth's whole mass of water to Ice and then, due to changing inclination reduces the ice to water as we are still experiencing today. The Dynamics of this phenomenon are a direct result of spiral planetary motion. We are today, in the last, the passive stage of that previously very violent process.

Due to the evidence, the annual cycles of freezing and melting is quite natural; no external forces or cataclysms are responsible for the ice age or the flooding. It is widely thought that the issue is one of 'Ice - Ages' when the reality of what we are examining is one Ice-Age, one cycle. Whether there other previous ice age cycles is another question altogether. When we consider that the point of Earth's travel at 90° is the apex of the water to ice cycle process. Then if we look at **Fig. B-8,** we can see that it is quite possible that the whole of **quadrant 3** is but the other half of our current ice age. This would explain the puzzling evidence of forests discovered beneath the ice in high latitudes as described above by geologist in the 1860s and also the fact that it has been recently discovered that an ice sheet once extended down to latitude 50° North in the region of the UK And that in that same region, now the North sea, the land mass (Doggerland) was once fully exposed because of the low sea level. Which signifies that there was a lot of water transferred to ice at that time. Causing territories such as Doggerland to once be high and dry when the ice ruled and the sea levels were lower earlier in our declining ice and rising sea level.

This information begs us to look a little closer at the

situation, because, if the regions found now still under ice in between Latitudes 65° and 75° North were once temperate latitudes then they were inundated beneath the ice long before Earth reached that point of 90° which was the apex of the development of the ice after which, the ice age began its decline.

Therefor the growth half of the ice age took place in quadrant 3 and this would explain how the polar regions were overwhelmed by Ice and only now, the once milder temperate forested regions are being exposed with the retreating ice caps.

It becomes obvious that in Earth's travel from 90° to 23°.5 the general process governed by the progress of its decreasing inclination is the reduction of ice to water. Whereupon, in the case of quadrant 3 the opposite was the case, the general movement was generating the reduction of water to ice. This would explain how afforested temperate regions around the parallel of 45° became buried in ice to this day. And also because of the reduction of ice releasing the water to increase the level of the sea, the previously dry landmasses are inundated.

It seems that they existed high and dry in the period of increasing ice and have been iced up back much further than the 90° peak when ice began to decrease

It will be realised that through quadrant 3 Earth's polarity is opposite to that in quadrant 4 and would have been spinning and orbiting in a retrograde fashion in relation to the Sun, and her poles and polarity reversed.

So here, we have now entered into a completely new conception not only of the ice age but of the solar system too.

Chapter 5: The Crux of the Matter

The Obliquity of the Ecliptic

Since the whole question of Earth's inclination rests on the interpretation of the *cause* of the annual change in the obliquity of the ecliptic. We need to examine by what means is it, that any change in the inclination of a spinning body takes place. We know that change in inclination takes place by the effect of an exterior force; in the case of the Earth, the principal effect is the Sun pulling on Earth's equatorial proturbence while there are secondary perturbing effects such as the Moon and other planets. This perturbing influence cannot alter the Earth's inclination but merely cause the Earth to waver slightly along its path, a perturbation as in the Moon's nutation effect.

The current official interpretation of the cause of the annual decrease in the obliquity of the ecliptic, developed and held since soon after Sir Isaac Newton's time is, that the perturbing effect of the planets prevents the progressive change in Earth's inclination to the Sun. This in spite of the fact that the Luni-solar effect is tending to progressively alter the inclination of Earth, by pulling on the equatorial bulge of Earth. This will equate the Earth's equatorial plane with the plane of the ecliptic as Newton defined, by causing a continuous decrease in the obliquity. The object of the official interpretation is to preserve the medieval notion of an 'invariable' Solar System and conceal any notion of a changing Solar System.

In order that an interested reader might understand how the deception to support the notion of the 'invariable' Solar system unfolded. Where the object was to avoid recognition of the *continuous* annual decrease in the obliquity of the ecliptic. I have included here, the fine detailed extract of the history of the creation of this medieval notion by George Dodwell. Beginning from an assertion that Newton first suggested the notion, on through the various exponents of the theme. Halley, Euler, Lagrange, Stockwell, and Laplace. How they each, in turn added to refine this alternative interpretation concluding in the establishment of the notion of the oscillating or 'wobbly Earth, as we see in the following excerpt by Dodwell:

History of the 'Wobbly Earth' Theory

An account of astronomy's development of the current conception of the cause of the decrease in the obliquity of the ecliptic by George Dodwell B.A., FRAS.

With acknowledgement to publishers website - Genesis Science Research. < http://setterfield.org>: Excerpt from Chapter 1:

THE OBLIQUITY OF THE ECLIPTIC

Ancient, mediaeval, and modern observations of the

obliquity by George. F. Dodwell of the Ecliptic, measuring the inclination of the Earth's axis, in ancient times and up to the present by:

George F. Dodwell B.A., FRAS.

INTERPRETATION OF THE CURVE

........The periodic movements of the earth's axis, however, are completely known; and there is no force available to cause any additional movement.

The principal movements of the earth's axis of rotation, in order of their discovery are:

(i) the luni-solar precession
(ii) nutation
(iii) the planetary precession

What is known as the "general precession" combines the luni-solar and planetary precessions. The luni-solar precession is a gyratory movement (like the wobble of a spinning top) of the earth's axis round a central point, the pole of the ecliptic, in an average period of 25,700 years, this period being lengthened to 25,800 years by the effect of planetary precession.

The earth's precessional movement was one of the great discoveries of Hipparchus, about 124 B.C.

A change in the obliquity of the ecliptic, dependent on the planetary precession, was first suspected by Eratosthenes in 230 B.C. As we have seen above, Eratosthenes found, from a long series of observations, that the obliquity was 23° 52' in his time, so that it was 8' less than the value 24° handed down from the time of Thales, in 558 B.C., and of Pythagoras, in 515 B.C.; i.e. a decrease of 8' in about 300 years.

Nearly 2.5 centuries later, about 14 A.D., at the end of the reign of the Emperor Augustus, the Roman mathematician Manilius noted a change in the obliquity of the ecliptic, which he attributed either, as he says, to "the discordant course of the sun itself, and some change in the sky, or through some change in the universal earth, by which it has moved away from its center, **as I have detected myself, and I hear of also in other places.**"

This remarkable conclusion was reached by Manilius from his observations, during 30 years, of the solar shadows at the summer and winter solstices, cast by the great obelisk in Rome. This great obelisk, 75 feet high, was provided by Manilius with a golden ball at its summit, and the position of the circular shadow on the flat pavement below was carefully measured by him, over a period of 30 years, with a brass scale fixed into the pavement.(7)

This movement of the earth, together with a supposed backward and forward movement of the equinoctial points, came to be associated in later years with theories of "trepidation," which was defined as "a motion ascribed to the firmament, to account for certain small changes in the position of the ecliptic and the stars."

Copernicus, in 1525 A.D., and Wendelin, the noted Belgian astronomer, nearly 100 years later, endeavored to explain the precession of the equinoxes, and the variation in the obliquity of the ecliptic, by trepidational theories.

These were not on a sound basis, however, and no physical principle could be found to account for the movements. The precession of the equinoxes was, in fact, inexplicable by astronomers for 1800 years, from the time of Hipparchus until the great discovery of universal attraction by Sir Isaac Newton.

In Newton's *Principia*, Book 3, Proposition 39, the problem of the earth's precessional movement was, for the first time, correctly explained and calculated. It was shown to be due to the gravitational attraction of the sun and moon on the earth's equatorial protuberance.

Two hundred years after Newton, Sir George Airy, the seventh Astronomer Royal of England, said that "if we might presume to select the part of the *Principia* which probably astonished and delighted and satisfied its readers more than any other, we should fix without hesitation on the explanation of the precession of the equinoxes,"

Newton's *Principia* was published in 1697; and 60 years afterwards, Bradley the third Astronomer Royal of England, completed his famous series of observations of zenith stars, lasting for a complete 19 year lunar cycle. This resulted in his discovery of "nutation," a periodical slight change in the rate of precession, accompanied by a 9.5-yearly change of 9" in the inclination of the earth's axis, on each side of its mean position. The complete period was thus 19 years.

Bradley's discovery was published in 1748, and in explaining it, he indicated the true cause, that the "nutation" or nodding of the earth's axis, was due to the variability of the moon's action on the earth's equatorial protuberance, on account of the moon moving in an orbit inclined 5° 9' to the plane of the Ecliptic.

In the following year the eminent French mathematician D'Alembert published a treatise, in which a rigorous mathematical analysis conclusively proved this to be correct, and, in addition, it gave complete confirmation of Newton's explanation of precession.

In Newton's time the gradual diminution of the Obliquity of the Ecliptic had not been fully established. Some astronomers, like Tycho Brahe, Riccioli, Gassendi and Flamsteed, believed that the discrepancies, from ancient times until their own, were due to errors of observation.

Halley, the second Astronomer Royal of England, also held this opinion. In the year in which Newton's *Principia* was published, 1697, Halley presented a paper to the Royal Society, comparing the solstitial altitudes of the sun observed at Nuremberg by Wurtzelbaur in 1686 with similar observations made at the same town by Bernard Walther 200 years earlier. In this paper he said: "from these observations it appears that the obliquity of the ecliptic has continued unaltered for these 200 years last past, that is to say, that the angle which the earth's axis makes with the plane of the Ecliptic, or orbit wherein she moves annually around the sun, has been without sensible change in all that time."

Newton, however, had shown in the *Principia*, in general terms, that the perturbing effect of the planets would produce, among other things, an alteration in the inclination of the plane in which any one planet moves. He did not make any calculations of the amount of this change, but in this he indicated the cause which, in the case of the earth, produces the secular variation of the obliquity of the Ecliptic.

The question of planetary perturbations thus began to attract the attention of astronomers, and in 1756 the Swiss mathematician, Euler, in a series of papers, showed that the effect of these perturbations on the earth would be not only to cause a slight movement of the precession in the opposite direction to that produced by the sun and the moon, but also to cause the obliquity of the Ecliptic to diminish about 48" in 100 years; not much more than 1" higher than the true value.

Nearly 30 years later, the great mathematician Lagrange showed that the Obliquity could not diminish indefinitely, but that the action of the planets could only cause small oscillations in the positions of the various orbits, so that, in the case of the earth, the obliquity of the Ecliptic was confined within small limits.

Laplace, in 1827, calculated the total range of this oscillation of the earth's orbit to be 3° 7' 30"; but a later calculation by Stockwell in 1873, using more accurate determinations of the masses of the planets, made the total range somewhat less, namely, 2° 37' 22"; the limits of the obliquity ranging from a minimum of 21° 58' 36" to a maximum of 24° 35' 38".

Laplace, in *The System of the World*, Vol. 2. p. 211, explained the variation in these words:

"If we refer to a fixed plane, the position of the orbit of the earth, and the motion of its axis of rotation, it will appear that the action of the sun, in consequence of the variations of the Ecliptic, will produce in this axis an oscillatory motion similar to the nutation, but with this difference, that the period of these vibrations being incomparably longer than that of the variations of the plane of the lunar orbit, the extent of the corresponding oscillation in the axis of the earth is much greater than in the nutation.

"The action of the moon produces in this same axis a similar oscillation, because the mean inclination of its orbit to that of the earth is constant.

"The displacement of the Ecliptic, by being combined with the action of the sun and moon upon the earth, produces upon its obliquity to the equator a very different variation from that which would arise from this change of position only; the entire extent of this variation would be, by this alteration of the Ecliptic, about 12 degrees, and the action of the sun and moon reduces it to about 3 degrees."

Stockwell also commented in a similar way on the inter-action of the sun, moon and planets, in producing the variation of the obliquity, as follows:

"Here we may mention a few among the many happy consequences which result from the spheroidal form of the earth.

"Were the earth a perfect sphere, there would be no precession or change of obliquity arising from the attraction of the sun and moon; the equinoctial circle would form an invariable plane in the heavens, about which the solar orbit would revolve with an inclination, varying to

49

the extent of twelve degrees, and a motion equal to the planetary precession of the equinoctial points.

"The sun, when at the solstices, would, at some periods of time, attain the declination of 29° 17' for many thousands of years; and again, at other periods, only to 17° 17'.

"The seasons would be subject to vicissitudes depending on the distance of the tropics from the equator, and the distribution of the solar light and heat on the surface of the earth would be so modified as essentially to change the character of its vegetation, and the distribution of its animal life.

"But the spheroidal form of the earth so modifies the secular changes in the relative positions of the equator and ecliptic, that the inequalities of precession and obliquity are reduced to less than one-quarter part of what they otherwise would be.

"The periods of the secular changes, which in the case of a spherical earth, would require nearly two millions of years to pass through a complete cycle of value, are now reduced to periods which vary between 26,000 and 53,000 years.

"The secular motions, which would take place in the case of a spherical earth, are so modified by the actual condition of the terrestrial globe, that changes in the position of the equinox and equator are now produced in a few centuries, which would otherwise require a period of many thousands of years.

"This consideration is of much importance in the investigations of the reputed antiquity and chronology of those ancient nations which attained proficiency in the science of astronomy, and the records of whose astronomical labors are the only remaining monument of a highly intellectual people, of whose existence every other trace has long since passed away."

Other eminent mathematicians, from the time of Euler, also made contributions to the theory of planetary perturbations, and of the precession movement; so that, before the end of the nineteenth century, the action of the sun, moon and planets upon the earth's axis of rotation became fully understood and accurately calculated.

To sum up, therefore, the external forces, acting upon the earth, and affecting its axis of rotation, are the gravitational attractions of the sun, moon and planets. Of these, the sun and moon cause the luni-solar precession, consisting of a very slow gyratory movement of the earth's axis round the pole of the Ecliptic as center, in an average period of 25,700 years, lengthened by planetary precession to 25,800 years.

The inclination of the axis, however, is maintained at an almost constant angle so far as changes of a relatively short period are concerned; the principal variation of this kind, viz., that of the 19-year lunar nutation, being only 9" on each side of the mean path of precession.

The attractions of the planets, combined with the sun and moon, however, cause a displacement in the plane of the Ecliptic, so that it oscillates with respect to the equator in a long period, varying from 26,000 years to 53,000 years, and this produces in the obliquity a series of maximum and minimum, having a total variation of 2° 37' 22", namely, from an absolute maximum of 24° 35' 38" down to an absolute minimum of 21° 58' 36", according to Stockwell's calculations.

There are no other external forces acting on the earth to produce any additional change in the obliquity of the Ecliptic; '

End of Quotation

Authors Note:

Professor Dodwell suggests that Sir Isaac Newton had initiated the conclusion that it is planetary perturbations that are responsible for the annual decrease in the obliquity:

'Newton, however, had shown in the Principia, in general terms, that the perturbing effect of the planets would produce, among other things, an alteration in the inclination of the plane in which any one planet moves. He did not make any calculations of the amount of this change, but in this he indicated the cause which, in the case of the earth, produces the secular variation of the obliquity of the Ecliptic.'

This is a presumption of course, since it really goes without saying that Newton new well enough that all bodies have a perturbing effect on each other but it is doubtful if he was suggesting then, that the planets were the cause of the decrease in the obliquity. Be that as it may, the objective was now being set, the real issue was to establish that the planetary perturbations were powerful enough to account for the annual decrease in obliquity. The science of astronomy could not accept that the precession motion of Earth's axis could be any thing other than a perpetual circle. A different, more acceptable reason than the spiral to explain the decrease in the ε had to be found.

Chapter 6: The History of Deception

Still Nurtured in Modern Astronomy

Having many years of attempting to bring the attention of astronomy organisations and influential individuals engaged in the field of astronomy, to the subject of the spiral motion of the planets in the solar system; I have generally speaking, been met with a wall of silence. Really it is no surprise that this attitude, particularly of those professionals engaged in astronomy do not care to respond to what is, a new theory. This is because the theory has the potential to overturn the false medieval notions concerning the motion of the planets of the solar system, which have been accepted by the science over the last 365 years,

Key members of the scientific community depend on the established criteria, employing their expertise in teaching and propagating the criteria to support their prestigious status and livelihood. It is Natural they don't wish to jeopardise the status quo. They do not wish to stand up and expose the deception and fallacies that exist in the science. This is not at all surprising because it is historically, a natural reflex to change throughout society. The most striking example, for us, is Sir Isaac Newton's negative reaction to where his own and Dr. Robert Hooke's findings in around 1689 were leading to, an exposure. A movement of bodies to a centre. A topic brought into the open by Dr. Hooke in his 1674 work,

'An Attempt to Prove the Motion of the Earth from Observations'.

Understanding this state of affairs existing in the official theory, I therefore have confidently directed my efforts to propagate the subject matter to my peers, the amateur astronomers around the world with their clear open minds, who are unafraid and enjoy the freedom to examine new fields of research. I must say that the effort on the Web has proved most promising with growing interest by free, unbiased amateur astronomers around the world.

How Did This Situation Come About?

When Dr. Robert Hooke approached him in 1679 with a modern notion of planetary motion, Newton, said he was not engaged in the natural sciences and was involved in other studies and was not aware of Hooke's publication, 'An Attempt to Prove the Motions of the Earth from Observations' Hooke, during the course of their correspondence produced a rough sketch of his conception a spiral motion toward a centre. Not a circular spiral but an elliptical spiral which he called an 'ellipsoidal' (elliptispiral). This was a crucial indicator of the course that the topic of planetary motion being discussed in the correspondence had taken.

Three leading questions reveal the reason why astronomy failed to take that leap forward in the period of 1674 – 84 and failed to contest the medieval notion of the 'invariable' solar system this left the investigation into planetary motion where it still rests today, at the point where Kepler advanced it to, in the early 17th century:

1. Why did sir Isaac Newton not endorse the spiral?

Why did Sir Isaac Newton in 1679, not fully embrace Dr. Robert Hooke's work 'Attempt to Prove the Motions of the Earth from Observations', which focused on a 'movement toward a centre', virtually a spiral motion? After Hooke's death Newton published his Principia but never pursued the topic of a movement toward a centre as Hooke had hoped, when approaching him in 1679. The Principia is a classic work on natural science but instead of pursuing the 'elliptisoidal' spiral theme of planetary motion, proffered by Hooke, he limited the study to 'curvilinear' motion the study of curvature, which clearly was a diversion that shifted attention away from the study of movement toward a centre. Yet, in spite of this he was bound to accept the ultimate conclusion of planetary movement toward a centre as he explained the motion of the moon in the Principia (book 1 Motion of Bodies, Basic Concepts, Definitions and concepts. Definition V. book I 1687) and subsequently had to find an answer to satisfy the establishment's view of a 'perpetual' Solar system'. This he most cleverly accomplished and escaped the wrath of the establishment and his own castigation with a ruse saying that it is up to future mathematicians to find 'the force' that retains the moon in her orbit preventing it to move closer to Earth.

2. Why did sir Isaac Newton omit the movement of earth's axis?

Later (1685) when sir Isaac Newton announced his observation that the Sun's torque on the Earth' protuberant equatorial bulge, also described as the 'bending moment'; tended to draw Earth's equatorial plane to equate with the plane of the ecliptic? It is noteworthy that he did not mention that the Earth's north pole and axis would at the same time, be closing with the pole and axis of the ecliptic. Obviously, this would naturally account for the annual 0"47 decrease of the obliquity of the ecliptic indicating spiral motion. Why then has Newton's omission never been questioned?

3. Why did sir Isaac Newton not recognize the moons spiral motion?

Again in 1687 when sir Isaac Newton recognised that the Moon was orbiting closer toward Earth, why did he then state that, mathematicians are to discover the 'force' that would 'retain' the moon in her orbit about Earth? Recognition of the mere fact that Moon is getting closer is enough to give recognition to spiral planetary motion.

There is a very good reason for the negative reaction to what was potentially a revolutionary discovery, as Hooke said in the above work,

' ... I have in some of my foregoing observations discovered new motions even in the Earth itself, which perhaps were never dreamt of before'

The nature of the 'new motions' was evidently ground breaking in terms of astronomy, both Newton and Hooke would be well aware of the serious repercussions to follow if it were pursued. Had not Galileo just 40 odd years earlier been forced to recant his Copernican views under the threat of torture? We may grant that Newton in 1679 it could have considered there was good reason not to expound on a new spiral account of planetary motion. It may have been considered that society was not ready for such a leap forward, so the deception was founded, the medieval 'invariable' solar system notion was secure, and the deviation was to succeed.

Although the motive of the principal figures, Sir Isaac Newton, and Dr. Hooke concerning the deception that was firmly established in Newton's 1685 'Principia', can be well understood. The same cannot be said of subsequent leading astronomers through to the present day, who have been progressively relieved of the same fear and pressure from the establishment. True, the pressure is still extant, the establishment as a whole, still holds the medieval view of planetary motion as sacred. Even though today, one perhaps, would not be threatened with torture or death to expose it.

We well understand the precession of the equinoxes induced as it is, by Luni-solar gravitation, resulting in the retrograde motion of earth's axis of rotation about the pole of the ecliptic. This is the true description of 'precession'; it defines the change of axial inclination between a primary and its satellite affected through the spin of the satellite by the primary's gravitational influence. It is reasonable to conclude that the Sun has this effect on all planets in the solar system, due to their equatorial protuberance, just as Isaac Newton indicated with Earth.

But with regard to the primary's influence in the case of the satellites orbital motion, the primary's gravitational affect results in advancing the angle of the orbital plane and its axis to equate the satellites orbit, with the primary's equatorial plane. The satellites direction of orbit being the same as the primary's prograde rotation. This is a progressive movement as opposed to the precession movement. The data for these two opposite movements constitute the so-called 'constants of precession' 50.3" and 0".1247 per annum, 50".3 minus 0".1247 = general precession.

Laplace Confusion on Planetary Perturbations

Pierre Laplace was instrumental in bolstering the notion of an invariable solar system; the whole of the current theory of an unchanging solar system rests on his work on the effect of the perturbations of the planets. We must be aware of his confusing use of the precession (retrograde) movement associated with a planets spin and the (prograde) movement associated with a planets orbit. They are two quite different motions and move in opposite directions,

In his work on planetary perturbations Laplace was fulfilling the need for a theory to support the 'invariable' solar system concept, he introduced the idea of the invariable plane around which, to construct an explanation as to why the annual 0".47 decrease in the obliquity of the ecliptic is not a natural process bringing the Earth's axis to equate with the axis of the ecliptic and revealing the eventual merging of the equatorial and ecliptic planes. He was obliged to account for the annual decrease of the distance of the earth's pole toward the pole of the ecliptic but had somehow to show that this decrease was merely a perturbation caused by our sister planets disturbing Earth's axis. In fact he invoked a perpetual oscillation (wobble) of Earth's axis to increase and then decrease the Earth' inclination by 2° to 3° around our present inclination of 23.5° as it proceeds along its 26,000-year precession path about the ecliptic.

Now, any sensible person examining this supposed wobble of Earth's axis, the pole of which it is purported to decrease and then increase its distance from the pole of the ecliptic. Will understand that any perturbations affecting the Earth's axis and pole as it progresses along its precession path about the pole of the ecliptic will cause it to move along in a spiral fashion about its mean path. This effect is witnessed for example in the perturbing effect of the Moon upon the Earth's path as the Moon orbits Earth (nutation). Applying the logic of nutation in the case of planetary perturbation effect upon Earth the effect is fundamentally the same, producing a spiral motion. There is no doubt that such planetary perturbations though slight, similarly affect Earth on its orbital path. Wherever they turn they trip over the universality of spiral planetary motion.

The adoption by astronomy of 'Laplace's invariable plane' theory serves to conceal the fact that the centre of mass lies close to the centre of the Sun and that the Sun is the governing influence on the motion of the planets and their satellites. All are being compelled by means of gravitation to move to a centre, directly and indirectly; the mode of transition is spiral motion, an inherent feature of all matter in motion. If there was really a need to define an invariable plane, then the equatorial plane of the Sun naturally performs the role of an 'invariable' plane; for the term of the life of the Solar system, it is an invariable pathway, the guiding influence for the planetary path to the Sun.

The Laplacian theory, by promoting the notion of planetary perturbations as the cause of the annual decrease in the obliquity of the ecliptic is an attempt to obliterate the truth, that the annual decrease in the obliquity is manifestly the evidence of the spiral movement of the earth's pole moving in a spiral manner closer to the pole of the ecliptic. The proponents of the Laplacian theory are oblivious of the fact that even their argument for a 'wobbly' cycling motion of Earth's axis, could only be performed by means of a spiral motion, the very evidence they wish to negate

It can be seen that inadvertently, they are giving recognition of spiral planetary motion because it is implicit in any wobbly description. It is quite clear it is considered that the cause of change in inclinations are *planetary perturbations*, in order to replace Newton's Luni-Solar description of the cause of precession which more obviously indicated spiral motion. Once spiral planetary motion is recognized by astronomy, for it cannot be denied, then it has to be accepted that the Earth's inclination has always been changing, this is the crux of the matter, and the choice is either, perpetual circular planetary motion or spiral planetary motion.

Chapter 7: For a New Astronomy

Self Interest

Referring to figure **A** in chapter 1, clearly demonstrates that any movement of an axis changing its inclination can only be accomplished through a spiral movement indicating that this is a law of planetary motion. While subtle influences such as planetary perturbations will only cause oscillations (a wavy motion) of the planet along its orbital path, just as the nutation effect of the moon already has upon Earth on its orbital path.

When the Earth orbits the Sun we know that the orbit is not a circle by the mere fact that on completion of each orbit it does not return to the same spot from where it started the orbit. Even a schoolchild will recognize this as a spiral orbit. The Earth advances its axial position annually by means of this spiral motion, we know it as the precession circle but it is clearly not a circle. So we must ask why haven't astronomers recognized the spiral nature of planetary orbits from the outset?

To answer this, we know it is common knowledge that Sir Isaac Newton proved what was previously thought, that the effect of the Sun on the equatorial protuberance of the spinning body of Earth, caused the advance of each of Earth's orbit. This was established as 'the precession of the equinoxes'. This was applauded and accepted but there was no mention or recognition of this motion being due to the orbit being a spiral motion.

Further to this, Newton said that due to Earth's equatorial protuberance, there was a tendency for the Sun to pull the Earth's equatorial plane to coincide with the plane of the ecliptic.

Again, there was no mention of the fact that, if in each orbit of Earth, the equatorial plane of Earth shifted slightly toward alignment with the plane of the ecliptic. Equally important was the fact that the Earth's pole with each orbit, also moved in unison toward coinciding with the pole of the ecliptic. So there was from the beginning, in 1684 with the publication of Newton's Principia, a deliberate intention of astronomers to ignore the fact that the Earth's orbit is a spiral.

The reason for this is simple enough, it is of course, self interest, Newton himself was a prominent member of the church, and knowing full well that the churches generally, held the view of the 'invariability' of the Earth and the solar system. To oppose this view was a dangerous proposition. Who would dare to suggest that solar system was constantly changing, albeit very slowly but changing never the less? Ever since, principal astronomers in their work, repeatedly stress the point that *"the Earth's equatorial pole and plane will never equate with the pole and equatorial plane of the ecliptic"*. This was viewed as necessary to secure your career in astronomy.

With the current situation in mind in the 1980s, I thought to broadcast the view that astronomy was obliged to recognize this medieval 'invariable' view of the Solar system as obsolete and rectify the situation. I have been engaged in this activity ever since, to no avail, I have always been met with a wall of silence. This has not surprised me at all because the reaction is historically consistent with the response to change and advancement. The principle factor of this professorial resistance is the self-interest solidly bound to the established medieval view.

This situation demands the establishment of a **new astronomy** based on the true facts of Spiral planetary motion.

Appendix 1: Open Letter and Appeal to All Geologists

First published in 2016

I wish to circulate this message in order to bring to notice, information of a geological and astronomical nature, which is of great import and interest to the science of Geology and which will advance on current geological endeavors to explain the cause of the ice ages and tropical vegetation in the northern latitudes etc. The main contention has quite correctly been, that the cause has been due to changes in Earth's inclination.

My own researches have led me to some startling information on the subject. But first allow me to introduce myself. I am 84 years of age, an amateur astronomer for all my adult life. Presently, I reside in Australia, having left the UK back in the 70s. My research has been on matter and motion, the Solar system and finally and specifically on planetary motion.

Having very recently successfully completed my research on planetary motion, to my surprise, I found my study had produced an added bonus. It had led me to a quite new and logical explanation of the cause of the ice ages. This may seem a rather fantastic claim and hard to believe but I assure you that it is quite true, well founded and supported with the proof of almost 2,000 years of recorded astronomical data. The key to the discovery is found, in the 300-year-old distorted interpretation concerning the historically recorded data of the obliquity of the ecliptic, which is directly linked to changes in earth's inclination.

The specific criteria to indicate the annual reduction in the inclination of the earth, is the recorded **annual decrease** of the obliquity of the ecliptic of 0" 47 (or **47"** per century). This annual decrease is caused by the **spiral** law of planetary motion. The decrease results from the pole the Earth while precessing about the pole of the ecliptic in 26,000 years at 20".8 per annum. Annually decreases the distance of the earth's pole to the pole of ecliptic, thus annually reducing the earth's inclination. (see <u>fig A</u> below)

Accordingly, with a newfound interest in the cause of the ice ages, I read much on the studies of men such as Agassiz and Mr. Croll, for example, whose contributions to the Transactions of the Geological Society of Glasgow. 1867. I found that his conclusions to explain the reason for the fossilized vegetation and forestation in even as high as 74° north latitude, amongst many other finds, were due to some considerable change in earth's inclination. But astronomy offered him, like others who have followed, no possible cause, other than a meager annual change to the obliquity such as the small Laplace theory of a 2°-- 3° oscillating backward and forward obliquity, centred about the current 23°5 obliquity

Thus, according to this theory, the earth's inclination is virtually unchanging; astronomy deals with the annual decrease in the obliquity of the ecliptic by giving the cause as, the perturbations of our sister planets producing the oscillating affect on the earth's axis. Thereby, this denies any *continuous* movement of Earth's pole moving as it is currently, toward the pole of the ecliptic at the annual rate of 0"47. It not anywhere acknowledged that any change in the axial inclination of any spinning sphere could only be accomplished by means of spiral motion.

Consequently geologists have been shackled to this distorted view of the earth's 'oscillating inclination'. Consequently, geological researchers seeking the cause of those dramatic changes in climate needed to produce the ice ages; naturally, had turned to astronomers to explain. The most obvious cause has always thought to be, changes in the inclination of earth. Since no sound explanation has been forthcoming from astronomers, in frustration, many theories have been propounded over the years, it is still generally regarded as a mystery.

This will continue to be so, until geology frees itself from those shackles. I speak from experience when I say any consideration or interest in the topic from the science of astronomy has been well tried without effect. Astronomy clings staunchly to the medieval view of an unchanging inclination of Earth. Today, the quest concerning the mystery of the ice ages is bound up and centred mainly on the popular 'Malenkov' theory. There are others of course, such as the newly emerging 'Snowball earth' but they are all still hedged in by the notion of a virtually unchanging inclination of planet earth. A solution was never to be found due to this situation.

Geology Needs to be Freed From the Farce

Sadly astronomy had long since let the science of geology down, by ignoring the true facts concerning the obliquity of the ecliptic. This state of affairs has continued to this day. Astronomers can get along comfortably with the situation as it is. Unfortunately the majority are focused on space, on the cosmos, on the trajectories of artificial satellites etc., most have eyes glued on telescopes. The motion of the planet they are standing on is of little interest to them even when it is suggested to them that something is amiss. This has certainly held back progress in the science of geology. However, I do believe that amateur astronomers are potentially more receptive to new ideas.

Geologists desperately need an explanation, it is vital to our understanding of what has caused the dramatic changes to our climate in the past and the changes to come. The reality of the situation is, that geologists will have to become familiar with the issue of the obliquity of the ecliptic, in order to advance their understanding of the continuing change in the Earth's inclination. This regardless of astronomy's indifference, the prospects concerning the Ice Ages are exciting; the real story stands before us. It is up to geologists to take the lead, to discuss and get involved. This includes the amateur geologists and astronomers who can have an important role to play. The facts and data await your scrutiny, astronomy has failed us all for an explanation of the cause of these phenomena, and meanwhile, the entire world continues to wait for the 'mystery' to be solved.

My life-long study has revealed that the annual *decrease* in the obliquity of the ecliptic is a constantly advancing component of planetary motion, and yes, the inclination of Earth has been greater than 23°.5 much greater. It is in the interests of all geologists to investigate this matter of the obliquity of the ecliptic, since no interest has been forthcoming from astronomers over 35 years of my approaches.

I draw attention again to those essential details of planetary motion that astronomy has historically ignored and distorted. The facts of the issue are that this has effectively obstructed any progress by geologists to find the cause of historical changes in earth's climate and the cause of the ice ages. For the whole story concerning the suppression of the role of the spiral in planetary motion, including the facts concerning the annual decrease in the obliquity of the ecliptic. It has been recently published in. **'The Dynamics of Spiral Planetary Motion'** available from Amazon books.

As previously stated said, the Ice Ages are a startling, awe inspiring phenomena. I am continuing my study on some finer details of the fascinating process of the ice ages.

In conclusion, it is my earnest desire and hope, that geologists will familiarize themselves with the facts outlined above concerning the obliquity of the ecliptic. This in order to enlighten the science of geology and advancing astronomy from the medieval notion perpetuated concerning the Earth's inclination and climate.

The simple fact of the matter remains: One either believes that the Earth's axis and equator is moving continually, by annual increments, toward alignment with the ecliptic and its pole, or believes the adherents of the notion that the Earth's axis is eternally oscillating (wobbling) about the present day inclination of the Earth's pole of 23 °.5. Simply, it's a question of believing in perpetual motion, or spiral motion.

Appendix 2: The Author's Tale, of a Tail

The Boy Who Chased and Caught a Spiral By the Tail

The boy, a mere 9 years old, was playing with water in the kitchen sink, looking with curiosity and wonder at a strange display of nature as the water drained from the sink. The water didn't fall down the plug - hole as he expected, it flowed in a particular way, it had a distinct order about it. No matter how often he returned to play with the water to observe the motion, it was unchanging as if by a law it always obeyed. It was of course a spiral motion.

As the boy grew up, he more and more became aware of this law operating in the world around him. It was there in all the natural sciences, he was to find that it operated in physics, atomic structures, genetics biology, chemistry, meteorology, astronomy, stellar and more.

His curiosity grew during his teen years but search as he may, though defined there was no scientific study of this universal form of motion. No recognition of the universality of the spiral in nature. Yet everyone is aware of this spiral phenomena that operates around us in those many magnitudes, though merely taken for granted it is left for the scientists to explain to us, isn't it?

So what he wondered, had the masters of mathematics physics and astronomy to say about this phenomena?

Then in good time he sought throughout history their views, from the ancients through to Copernicus, Galileo, Kepler and Newton, also those others who followed. But no! There was no particular study or recognition of this natural phenomena of the spiral as a universal law. Sir Isaac Newton's definition of the motion of matter was accepted as 'mere change of place'. A study of the movement of circles and curvature in orbits is all that has followed.

This was a severe case of scientific neglect, the now young man thought; evidently nature's poor spiral had no place in our sciences, especially strange since spiral motion was patently unavoidable. For example, spiral motion has to be accounted for in every single orbit of man's artificial satellites. They could hardly get anywhere with circles could they? But the spiral is still taken for granted and treated as being of little consequence. Had they not mastered it by mathematical computation to suit the investigation of space with rocket science?

Though our man without the benefit of higher education did have the simple power of observation. Recalling for example, his first observation as a boy when he gazed up at our beautiful companion in space the Moon. And who cannot be enthralled at her wonder, beauty and serenity as she rides our sky? What holds her there he thought? A tantalizing question for him as his curiosity grew. She *is* moving around us and as the physicists explained, there is no such thing as a straight line or a closed circle in nature. These are fundamental scientific observations, which are our guide to any investigation in physics or astronomy. As is the declared fact that science recognizes no such thing as perpetual motion.

Thus the obvious explanation of our companion's motion about Earth is that, she is traveling in a spiral, it is undeniable but the question then arose was, is it closing with Earth or receding from her? This brought in the question of gravity, Isaac Newton's considered opinion was found to be that the Moon was coming closer to Earth *but* he declared, it is for "future" mathematicians to find the " force" that would check and hold the Moon in her orbit. He had tactfully chosen not to commit himself further on that question. He knew where any further study of spiral orbital motion would lead and his fortunes would definitely not benefit from it.

Surprised that science considered that the Moon should be *held suspended* in her orbit, our young student would later explore into the reason why? Instinct told him that regardless, of man's fears or will, gravity would not *repel* the Moon from Earth.

Subsequent mathematicians were to provide the force to 'prove' that this repulsion of the Moon was taking place and then even more like theories abounded. The Moon had to be mathematically stopped in her spiral course a very complicated process to be sure, requiring the most advanced mathematicians. Though in this day and age, it is simple enough for all laymen to understand

At this juncture came the realisation to our young enquirer that the spiral was an integral law of nature that governs the motion of matter. To begin with, it was so obvious a natural function of the spheres in the solar system and indeed further investigation confirms that the spiral is the *form of motion* of all spheres in creation. The fact of the matter is that the science of astronomy had really lost the plot on planetary motion and could advance no further.

The young man now knew that he held a virtual jewel in his hand that the masters spurned, he wanted to share it with the world but disappointment followed year after year, all approaches to the academies from top to bottom including astronomical societies around the world were unwilling to discuss the spiral governing the solar system and treated the subject with a wall of silence.

To conclude this very brief outline of his true story, eventually our inquisitive man of no great learning fully investigated the subject himself. It was a lonely but very exciting path over many years with many discoveries unfolding, explaining many of the remaining mysteries and history concerning the development of the solar system, ranging from Earth's growth to climate change, to the only logical explanation of that great mystery of the ice-ages. Now this man is old, but quite content and satisfied, having lived a normal industrious life, rich in the knowledge to have caught and followed the spiral in its course, to unlock a real natural treasure trove to be sure.

See, he's let go the spiral's tail, now its free and for all to use and explore.

Alan Dover Oct. 2016

References:

Astronomical Almanacs - The historically recorded astronomical Constants.

Data on the motion of the Earth and satellites in the Solar system.

The History of the principal Astronomers mis-interpretations of the annual decrease in the obliquity of the ecliptic from 17th century onward.

The scientific principles - that in nature, there is no such thing as perpetual motion, and no such thing as a closed circle or such thing as a straight line

Observations of spiral motion in the solar system and throughout nature.

'The Dynamics of Spiral Planetary Motion' 2nd edition. Amazon books.

Dr. Robert Hooke. 1674 'An Attempt to Prove the Motion of the Earth from Observations'.

Isaac Newton. 'Principia' 1687. Book 1. The Motion of Bodies'

Lieut. – Col Drayson. R.A. F.R.A.S. works -1874.

The Origin and Decline of the 'Ice Ages'